STEM CELLS

USING THE BODIES OWN CELLS TO TREAT
INJURIES,
REVERSE AGING
AND NOW
REGROW HAIR

JOHN SATINO

ISBN: Softcover   978-1-5434-4484-1
        EBook   978-1-5434-4485-8

Print information available on the last page.

Rev. date: 08/23/2017

To order additional copies of this book, contact:
Xlibris
1-888-795-4274
www.Xlibris.com
Orders@Xlibris.com

FOREWORD

Since the beginning of time, we have been looking for the Fountain of Youth, this has opened the doors for charlatans and con men, thought out the centuries, to separate us from our gold and silver, with promises of eternal youth. The prime example in the eighteen hundreds was the Medicine Wagons traveling across the parries, offering entertainment, and cures for everything from gout to baldness.

Amazingly, the first DNA was separated from a human cell by Fredrick Meisner in 1879 now, many videos on Google and YouTube are showing eye opening treatments using Stem Cells to treat everything from Burns to Growing Complete Body Parts.

In my opinion, reversing the signs of aging is just around the corner. This is not an overnight revelation. Studies have been going on for years to get us to this point. The first public awareness of Stem Cell research started coming to view about nineteen hundred seventy nine. At this time several (in-vitro fertilization embryonic Egg Banks.) were being opened at major university research centers. This new technology would allow an embryonic egg from a donor to be fertilized in the laboratory, by a male sperm donor, then incubated. This was a great breakthrough for infertile couples, or for those planning future families.

The frozen eggs had to be stored, and kind of like your grandma's old fur coat, the cost of storage over time, became less of a priority. As many families discontinued payments for storage, these eggs were scheduled to be destroyed. Some researchers felt this was a great opportunity, to study these pure embryonic cells.

A politician, looking for recognition, as they all do, took this information to the religious right, then led them to believe, that these researchers are growing humans. This became a big issue, with government stepping in to ban Stem Cell Research, not just embryonic but adult stem cells were placed into that category, by a mislead or uneducated media.

Many researchers, who had already invested much personal time, left the US. So they could continue their work. This gave the European theaters great advancement over US scientist in the nineteen eighties.

When President Obama came in, at the behest of Several Hollywood types, and scientific advisers, he lifted the ban on Embryonic Stem Cell Research in the US.

The FDA also agreed to allow for Minimal Manipulation of Autologus Cells (This would allow us to concentrate, on a four to one basis) adult cells from Blood or Fat Tissue known as Platelet Rich Plasma.

Now the average Doctor or Clinician could start to use some of this science to offer some limited treatment to their patients. Many Doctors today are using this technology to treat Injuries, Joints, Backs, Knees and so on. At this time Insurance Companies will not pay for this service they consider it experimental medicine, they, in most cases, will pay for a cortisone shot in the same area, which makes little sense.

THE LASER CONECTION

My first exposure to Lasers was at NASA in nineteen eighty two & eighty three. I was working as a consultant to the Bio Medical Department under direction of Dr. Paul Bucannon. I had purposed Electronic Muscle Stimulation for preventing leg muscle atrophy in space. Since the muscles are large they atrophy very quickly in a no gravity situation. The Russian's were using this technology in their space program, so I purposed it, using a small battery operating on a one fifth volt, through a carbonized rubber pad, placed into the space suite, for delivery. We simply had to find the motor point of the leg muscle in each astronaut, then place the pad in the space suite. Then just connect the wires.

While there I was introduced to Dr. Mary Ann Frye. She was working on, among other things, Low Level Lasers. This Cold Laser of approximately 650nm can penetrate tissue, reintroduce vascular systems, and accelerate cellular division, with no physical sensation. This was purposed to treat a space induced injury Due to the lack of gravity, space injuries do not heal as under the Earth's gravitational influence.

Jumping head to two thousand and three, Dr Mike Markou and I conducted the first US Clinical study using this Low Level Laser to treat Hair Loss, Published in The International Journal of Cosmetic Surgery and Atheistic Dermatology, the findings of this study lead to future FDA clearance of this technology to treat hair loss.

Then in two thousand eleven Professor Uri Oron in Tel Aviv discovered this Low Level Laser could be used to re-activate Stem Cells in vitro. A demonstration of a heart attack victim's bone marrow being exposed to Low Level Laser caused an eighty percent reduction in formation of scar tissue, around the damaged heart. This laser causes more proliferation in the bone marrow. I regularly use Low Level Laser to insure my increased Adult Stem Cell Count, and feel it is one of the greatest scientific breakthroughs of the Twentieth Century.

Hair Loss & Self Esteem

Hair loss was devastating for me, I came from an era of Brylcream, Fonzie and Duck Tail hair styles. Unlike today, you could not shave your head unless you were completely weird. My uncle worked at the state prison and told me they shaved the inmates heads once per week. This was supposedly for hygiene, but in case of an escape, prisoners with that hair style would be very recognizable. Since I didn't want to loose my hair or have a shaved head, this started my life long quest for hair. My earliest research lead me to Dr. Norman Orientrich, A New York Dermatologist that was rumored to be performing hair transplants on celebrities like Frank Sinatra. The method was crude, bloody and painful but worked. In 1953 Dr. Orientrich used a biopsy punch to remove 4ml. round plugs of hair from the back and sides of the head, or (donor hair). Then similar plugs of balding scalp would be removed and replaced with the donor hair. This works on the principal of donor dominance, since the hair from the back and sides is not effected genetically in the same manor as the hair on top, it usually grows for life, and acts accordingly when surgically moved to a new location. Due to the large size of these plugs, the appearance was somewhat like a Barbie Doll or Corn Rows. Now the procedure has greatly improved with microscopic surgery, most hair transplants are undetectable.

We now use a microscope to dissect the hair follicles into one two or three grafts and use a co2 laser to provide a Quick, Virtually Painless procedure with little or no downtime. There are however disadvantages to hair transplantation, besides the cost. You are robbing Peter to pay Paul. If the hair loss is severe and the bald area large, there may not be adequate hair from the sides and back to complete the job. In the 1960's this was a very expensive operation, and as a college student it was way beyond my humble means.

Hair Pieces, Wigs & Weaves

That is when I started to look at hair replacement centers in other states. They took hair samples, matched the hair in color, texture, curl and est. and magically attached it into my existing hair and for only a couple of thousand dollars. (I was sold). This was my introduction into a hair piece or hair weave. I didn't like the constant maintenance, and it was hard to keep the hair matching my own curl and color, but I had hair and could go on with my life.

Getting into the Business of Hair Replacement

At the time I was studying Biomedical Engineering, and working for a heart pacemaker company. But because of my own situation, I wanted into the Hair Business. I flew to Florida to meet the president of the Hair Replacement Centers. I immediately went to work setting up Hair Replacement Centers across the country, In the 1970's I was setting up one or two centers per week, which provided me with a greater income than I could have ever imagined. But after a year or so living on airplanes and hotel rooms, I wanted to open my own clinic and get off the road.

By this time Hair Transplants were getting more sophisticated, some doctors were using mechanical drills to remove the donor hair and could do thirty or forty grafts in one session, today we do two or three thousand at one time. I wanted my own growing hair and my research lead me to Dr. C.P. Chambers in Atlanta. Not only did he do my first hair transplant, but started to operate out of my Ohio Clinic two or three times per month.

I will never forget the feeling when I was able to take off my hair replacement and have the wind blow through my own growing hair, it was like getting out of prison. I was no longer a slave to my hair replacement, no weaves, glue, tape or potential embarrassment. I was now able to go to a regular barber for hair cuts just like everyone else.

Now that I had my own hair back, I wanted to cure baldness.

Working at NASA

But my life took a different turn in the 1980's, Through a friend that worked at NASA I found they were having problems with leg muscle atrophy, these large muscles atrophy very quickly in space due to weightlessness, a similar problem the Russians had within their space program in earlier years. During my work in heart pacemakers I had found studies coming out of the eastern block countries, using electronic muscle stimulators, initially for strength training, but also rumored to have been used in the space program. I had been using a sixteen channel EMS machine personally, so I forwarded this information on to Dr. Paul Buchannan, head of BioMedical Research at NASA. I was invited to come to the now Kennedy Space Center, to meet Dr. Buchannan and his team. After clearing security I was taken to a room full of doctors and engineers, I didn't answer any more questions than I had to, but had Dr. Buchannan roll up his pants leg, then attached six electrodes from the bottom to the top of his calf, this started his leg kicking wildly as I turned up the controls.

That was my introduction to NASA as a consultant in 1982. The use of a carbonized rubber electrode pads made this technology perfect for space suit applications. This allowed us to send one fifth of a volt into the motor point of the muscle, to cause the leg to contract at sub- minimal levels and prevent atrophy. The whole system was controlled and operated by a 9volt battery and weighted only ounces.

One other problem that existed was the inability to treat wounds in space, due to the lack of gravity and sun light. This problem was solved by the innovation of a low level 650nm cold laser. Similar to the size and weight of a flashlight, this technology could greatly decrease healing times if needed. But more about this later.

The First Drug Company Hair Research

In 1983 I got a call from my old friend and business associate Dr. C.P.Chambers. He had been in contact with Upjohn Pharmaceutical about a drug that seemed to be growing hair. He ask me if I wanted to be involved with this study, Of course (a drug that grows hair,) I couldn't wait to get to Orlando to start recruiting patients for this adventure.

The drug known then as Minoxdil or Lonitin was a smooth- muscle vaso- dilator blood pressure drug. It was nicked named the Werewolf drug because of facial hair growth.

There are thousands of small colorless hair follicles all over the face, the capillaries that feed the hairs are very small, When the drug relaxed these capillaries they filled with blood and started the small hairs growing in darker and thicker. Thus the Werewolf effect. Upjohn wanted to find out if applied topically, could this drug regenerate hair growth. So they turned to the top hair transplant and hair research doctors in the country at that time.

I worked as a research coordinator on this program with Dr. Chambers for two years. The FDA cleared version of the drug is known as Rogaine 5% for men 2% for women. This was the first FDA Cleared topical medication that had shown any validity in the treatment of genetically induced hair loss. Many are confused when they read the label, (for hair loss in crown or vortex only). Not that it doesn't work in the front, but the study's were done only in the crown for practical reasons, and that's the information received by the FDA, and so the approval is for the crown only. The original Rogaine was mostly alcohol base and caused a high number of patients to experience scalp problems. The new foam is a better carrier and has less scalp issues. Not a miracle drug but considered to be the first line of defense when starting to see hair loss. The drug must be used continually to get desired results, since once you quit using it, the capillaries go back to the previous atrophy state and hair loss can reoccur.

Merck Research on Propecia

By 1989 There was a new player on the horizon, Finasteride or Proscar a prostate drug developed by Merck. The male prostate enlarges with age, due to the conversion of DHT in the gland. This DHT also happens in the in hair. Proscar reduced the conversion by 65% and that allows the prostate to shrink to a more normal operation with increased urine flow. But men that took the drug noticed their hair was getting thicker and in some cases starting to grow back.

I was contacted by Merck to be research coordinator for this new study, in the Locations of Orlando, Tampa and Saint Petersburg Fl. In this double blind, placebo controlled study, Scalp biopsies were performed as well, to microscopically document the results. By the late 1990's a 1ml. version of the drug (Propecia) was FDA cleared as the first oral medication to treat hair loss in men with Male Pattern Baldness. Since this drug works on the root cause of male pattern baldness, It in my opinion, it is much more effective than Minoxidil, however the two, Rogaine and Propecia do seem to work better in combination.

New Surgical Laser Technology

By 1996 Dr. Chambers and I were starting to incorporate the co2 laser into our Hair Transplant programs. The initial work did not fair well. The non pulsed straight beam laser would virtually burn a hole, It left a clean dry surgical field, but caused thermal damage around the graft site effecting the growth of the new hair. Even though we had hundreds of thousands of dollars invested, both Dr. Chambers and I became frustrated with the outcome and went back o the conventional hair transplant methods.

Then in a shocking event took place over a labor day weekend, returning from performing hair transplant surgeries in St. Lewis, Dr. Chambers, his pilot, and medical assistants were all killed as their plane ran out of fuel on the approached to the Ft. Lauderdale airport.

After the loss of my long time friend and business associate, I wanted to make his dream of a laser assisted hair transplant a reality. By this time several laser company's had developed co2 super pulsed fiber optic lasers. Computer Controlled and very precise, we could now remove only the top layers of skin in the balding area, without thermal complications or damaging existing hair. This also helped to form new vascular systems to support the new hair growth. The laser would assist in the removal of the donor hair, with out the bleeding and trauma. normally associated with the old conventional hair transplants. This new laser technology now allows for a shortened procedure, with reduced healing times and far superior results.

I am always asked why doesn't everyone use the laser. For many years our competitors would simply quote old non-pulsed laser information, saying it effected hair growth or damaged the recipient area.

But the real truth is cost and learning curve. It takes extra effort and training time to become proficient, The lasers cost hundreds of thousands of dollars, not counting repairs and keeping the equipment maintained, The fiber optic is changed out on regular basis, and the laser tips are expensive and disposable. Compared to a conventional surgery using a thirty cent scalpel, The laser is not going to be the most profitable. But I do not believe there is a surgeon that can use a scalpel, with the precision or benefit of this new generation of lasers.

Low level laser treatment

In 1998 A new technology was being introduced in the US to treat hair loss. Low Level Lasers, not the lasers we were using in surgery, but a Cold Laser a light beam only with no heat, side effects or dangers. This was the 650nm low light laser similar to the one used for wound healing at NASA. This hand held laser came in from Australia, where it had been used for treating hair loss for several years. Now the developers wanted to break into the US market with their device. I met with David Michaels, President of Lexington International, His interest was only in sales and marketing the equipment, but he also understood the only way he could make any beneficial claims was to have Medical or FDA clearance.

I agreed to set up a pre-trial in my clinic to test the effectiveness of this laser not only for hair growth, but to treat patients after hair transplants to monitor healing times.

I would provide a laser immediately after surgery, and instruct the patient to use it only on one side of the head, and not to inform me which side. After one week they returned, a few started to see the results in a day or so and used the laser all over, but the one's that were compliant showed a vast reduction of healing times on the laser side. We published this information in various hair transplant and medical journals. I then contacted Many Hair Transplant Doctors across the country and sent them a laser, to conduct their own similar studies.

Dr. Michael Markou and I, set up a study to document the effects of this low level laser on hair growth, It started with thirty five subjects, men and women. A cast was used to count the hair in one centimeter, then samples were taken to test the strength and elasticity using a hair o meter. At three months, six month and one year the readings were repeated. Not surprising almost all subjects had some improvement, and many saw actual hair growth.

We reported the study's findings in the peer review journal, International Journal of Cosmetic Surgery and Aesthetic Dermatology 2003. After these articles Lexington International sponsored additional studies, then petitioned the FDA for clearance for the first medical devise proven to grow hair in the US. The FDA cleared the hair max in 2007.

PRP: The Future

This brings us to what I believe is the Future of Medicine (Adult Stem Cells)

By 2009 The FDA granted clearance for minimal manipulation of autologus blood for soft tissue injection. This means we can draw a patients blood, separate the platelets, that contain these healing cells and growth factors, then concentrate them four to one. Several safety guidelines were imposed, such as not transporting the blood over forty feet from the extraction site, or going to another floor to perform the injections. At this time there was only one manufacture of lab equipment cleared to separate these platelets, but now there are several others.

Working in hair replacement we constantly come into contact with patients that have diseases that can cause hair loss. Cicatricial Alopecia, this is a form of scaring of the hair follicle, that effects African Americans. Lupis an autoimmune disease and Alopecia Areata another autoimmune disease, some patients with this affliction wake up in the morning to find their hair on the pillow, with large completely bald spots on their scalp. This is usually devastating and confusing. Those that seek medical advice find there is no cure and little that can be done. Sometime doctors will inject steroids, but the results in most cases are less than favorable.

We had just such a patient, His hair loss was almost completely all over the scalp, with scattered clumps of hair in some areas. To minimize the unusual appearance, he was shaving the total scalp and wearing hats. He had been through the painful steroid injections form other doctors in past years, but nothing helped. He had stopped dating and turned almost into a recluse. We wanted to try the injections, but because of the extremely large area, and the long period of time the hair had been effected, we were apprehensive. But the patient convinced us he would try anything to solve this problem. On the day we injected his scalp, we placed him under nitrous gas and a local anesthesia to reduce the sting of the injections. He had also agreed to have our local TV news cameras there to record this.

Looking back this could have been a disaster, We were doing this on the six o clock news. Our reputation was on the line, if nothing happened, we would look like total quacks. But by one month we were starting to see results, three months a major difference and at one year his hair had completely grown back, and his life has changed, He now dates and has started a new job in health care. It has been several years and his hair is still growing. Since then we have injected over seven hundred patients, and use the cells in concert with our hair transplants. If an area is completely bald and has been bald for many years, there is probably not a live hair follicle or root structure. Hair is an organ, not as complicated as your heart or liver but an organ just the same. Once that organ is dead, we cannot create life. We have been using these growth factors to regenerate hair that is still in existence, even though it is fine and thin, there is a chance it may respond.

But there are no guarantees, the transplanted hair will always grow, and these cells may help maintain the existing hair. I have been in the hair replacement field for almost forty years, I have seen great strides in cosmetic medicine, and have been privileged to work with some of the foremost doctors and medical researchers in this field, I have not cured Male Pattern Baldness Yet. But I can see the light at the end of the tunnel, and it is not a train.

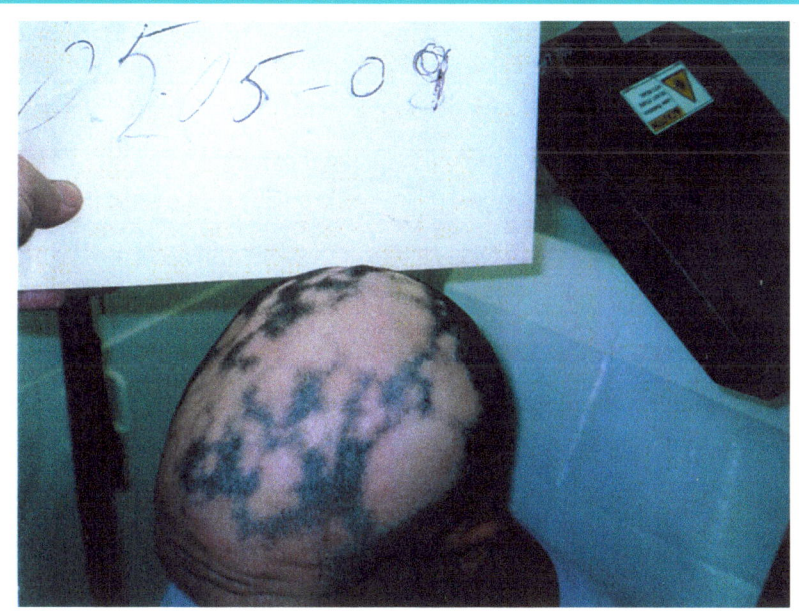

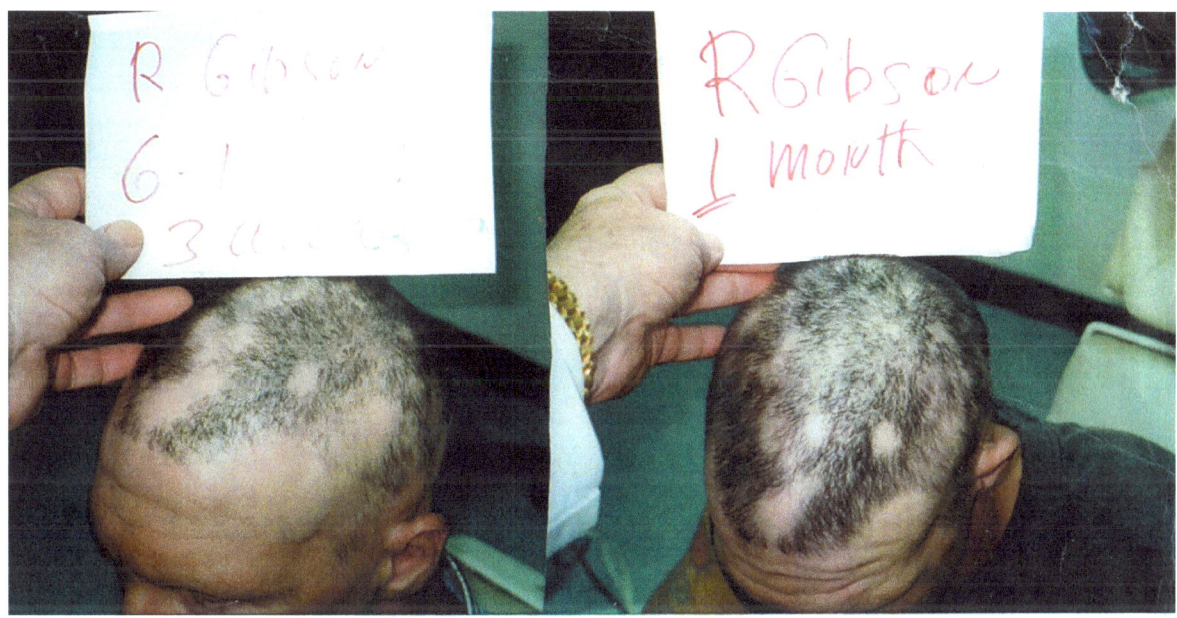

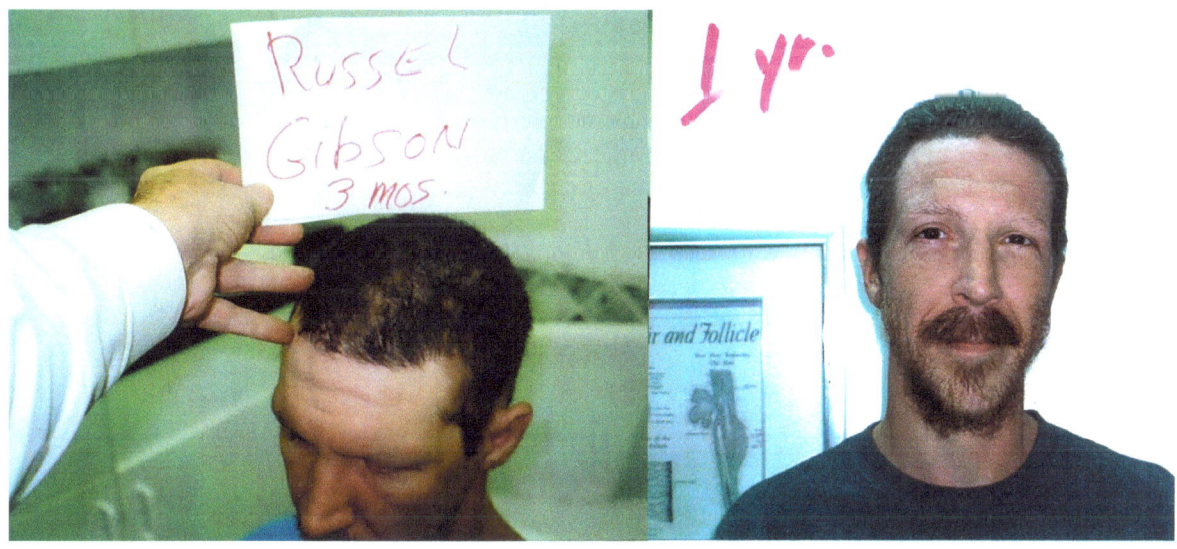

My Stem Cell Facial Restoration

In 2012 I found a limited number of doctors were starting to embrace this new technology for Aesthetics of the face and skin. After all the skin is the largest organ of the body, and potentially could benefit the most from this process. At my age I had the typical deep lines, folds and wrinkles associated with someone in their seventy's. After all I had my hair back for many years, Why not try to get a more youthful appearance. I had been taking lots of supplements, and one in particular to help release adult stem cells from my bone marrow. So I felt like a man twenty years my junior. But was starting to look my chronological age. The first process was the blood draw and preparation of the cells to be injected. I have never been fond of needles, so I used the nitrous gas, but found I didn't need it. Two small injections into the gum line, numbs the whole nose and mouth area. The first injections were of Hyaluronic acid, this naturally occurs in the body and is used in the deep folds and wrinkles to act as a scaffolding for the cells. This gives an immediate appearance change, but will be absorbed by the body in a few months. This is where the cells can start to form as a more long lasting substitute. The bones of the face start to degenerate with age as well, that's when you start to see the sunken look, by adding cells into that area it makes a major difference. Although the improvement is instantaneous, the formation of elastin and cell replication takes more time to occur. The final stage is one of the most important, the face is bathed in these cells then a derma roller is used to impregnate the skin all over the face. A collagen mask is applied for about twenty minutes to further drive these cells into the skin. When the mask is removed the skin looks remarkable. And as time goes on this gets even better as the body starts bringing in its own cells to the area.

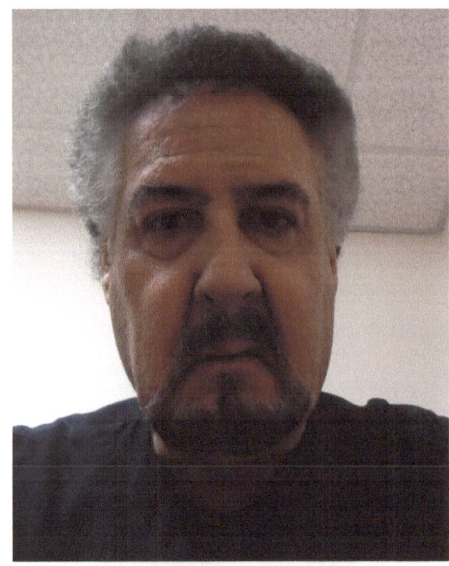

BEFORE

AFTER

After this, my greatest reward was going to my fifty five year high school reunion. I got a spray on tan, had my dentist do a bleaching and dressed accordingly. Most of my classmates looked like they should be in a Home. But it made me feel super to have everyone ask me, what are you doing? You look great. I just smiled and said, I just have good genes. No one knows where the future in stem cell research will go. As for now, we can only minimally manipulate these cells at a four to one ratio according to FDA rules. We cannot treat, multiply, or change these cells in any way in the US. But as time goes on I am sure we will have even better methods of harvesting and redistributing adult stem cells in all areas of the body. It is very exciting for me to know we are now living in a scientific world that only a few years ago would have been just science fiction.

VITAMINS & SUPPLEMENTS

Most do not get the recommended amount of Vitamin D, and one to two thousand IU's per day may be necessary, as most of us are low, even though we may spend a lot of time in the sun. Trace Elements and B Vitamins like Biotin may help with Cell Sub-Division. But massive amounts of vitamins like A, may actually be harmful. Non-water soluble vitamins should be taken in moderation.

SCALP HYGIENE (SHAMPOOS, CONDITIONERS, AND VITAMINS)

Shampoos, whether expensive or cheap, are basically soap. But Madison Avenue has done a fantastic job of selling us on these Miracle Elixirs. Of course there are different PH Balances, not that it makes a lot of difference once you get past medicated shampoos.

Conditioners simply make the hair slick so it doesn't tangle. Some of these products have coatings to make the hair feel thicker. These coatings usually fall into two categories, Animal Protein (Chicken Fat) or Polymer (Plastic). Yet, nothing will penetrate the hair shaft. The hair and fingernails are basically the same dead proteins stacked together. For anything to affect growth, it must take place in the root, not with a cosmetic thickener or coating. Natural cleansers and conditioners make sense because they contain the least amount of chemicals.

After researching these products for years, I feel the best results are found with a natural shampoo containing Emu Oil and Aloe-Plant extract conditioner, with Emu Oil.

The Boston University School of Medicine conducted several studies on emu oil and hair loss. Astonishingly, groups of men treated with emu oil versus corn oil had a 20% increase in hair growth activity. This was a blinded crossover study. So how do you get someone to use emu oil every day? I simply put it in the natural shampoo and conditioner.

DNA HAIR REGENERATION (Tomorrow's Technology Today)

Medical Device Companies are now using 3D printers to create Implants, Organs, and Prosthetics, Carbon Based Protein's with Human Blood Plasma are the key factors allowing this New Technology.

As Published in the Prestigious Medical Journal PUB-MED Abstract (Nanotechnol June ;3785-91)

Autologous Platelet Rich Plasma involes a simple blood draw. Then our Lab can Separate the Stem Cells and Growth Factors from your own blood. This Plasma is added to the Nanochrystalline Protein, then injected into the (Veli Hairs) thoughout the scalp.

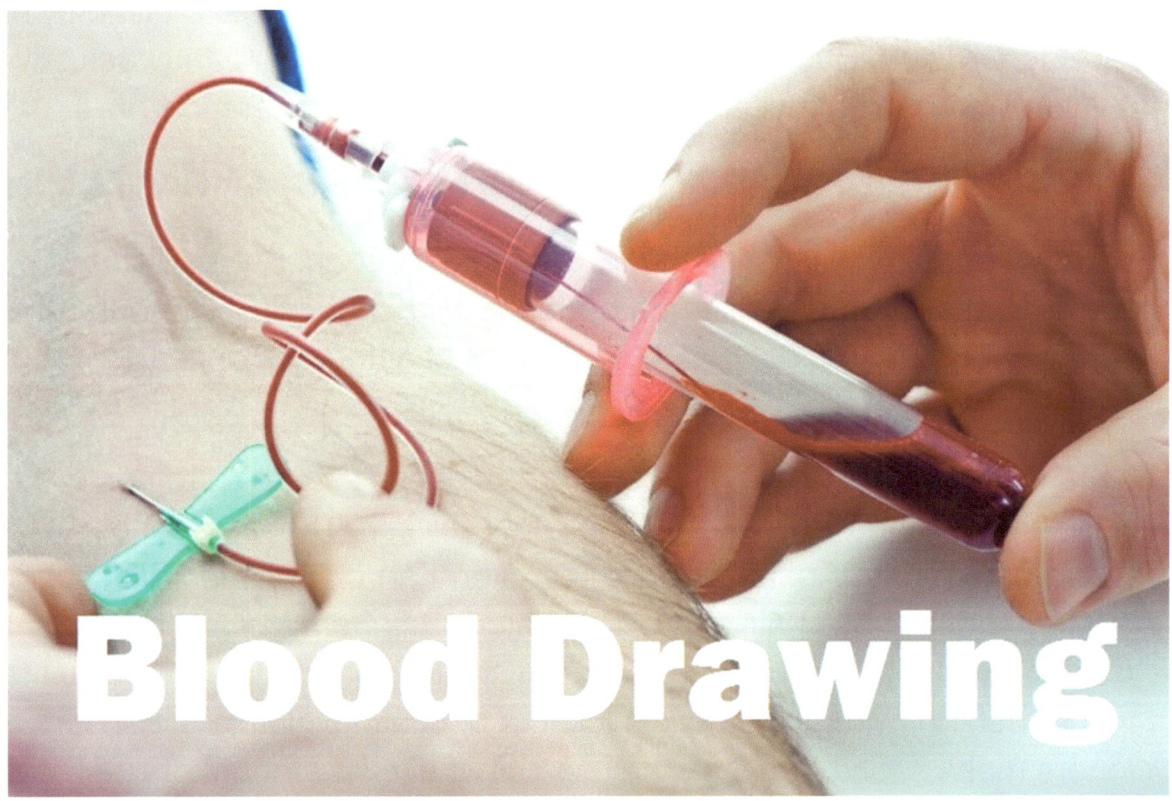

This Technology has proven to stimulate hair growth naturally within a few weeks, while immediately improving Density, Color, and Thickness.

A Topical Anesthetic Cream numbs the scalp as each hair injected, this insures a virtually Painless Procedure. Best of all, Visible Results can be seen as each hair is injected.

Hair Transplants are the only Guaranteed Way to grow hair in a Bald Area, But achieving density or thickness can be difficult and expensive. By adding this new technology to existing old hair transplants, we are now able to provide a Quick, Safe, Guaranteed Result.

Almost all bald areas have very thin, clear, short hairs. These colorless hairs show up under a microscopic exam but usually do not grow. When these thousands of hairs are injected, the results are a full thick looking head of hair, almost immediately.

We can treat Most Scalps that now have no visible hair, Beards, Sideburns and even Eyebrows.

All at a Small Fraction of the cost, of other options. No Surgery, Wigs, Medications or Side Effects, Just using your body's own cells,to regenerate your own hair. This is Today's Science, not science fiction.(see Before & After Photos)

A no cost Microscopic Scalp Exam is required to ascertain the number of Veli Hairs present in your scalp. This will determine how effective this process can be, in each individual case. In some cases photos can be relied on for time required and pricing.

HUMAN PLASMA PROTEIN COMBINED WITH PLATELET RICH PLASMA

In a study published in PUB-MED by Dr. Vinante and others, Plasma protein adsorption Into a nanocrystalline graphite(NG) with short time exposure to Platelet Poor Plasma show good anti-trombogenic properties This was detectable through fluorescent signal.

I feel this type of research becomes more valuable, as we progress into areas such as 3 D printing of Human Body Parts and Development of New Atologus Derived Therapies.

In a scenario of directly injecting each hair root, with protein enriched Platelet Rich Plasma, This could show much promise, in stimulating dormant hair follicles.

STUDIES OF (APHANIZOMENON FLOS-AQUAE)

As related to modulation of CXCR4 expression by an L-Selectin Ligand. Reported in the Prestigious Official Journal of (CARDIOVASCULAR REVASCULARIZATION MEDICINE) Volume 8 July –September 2007.

This study showed a 18% to 25% increase in numbers of circulating CD34 stem cells, maximized 1 hour after consumption of (Aphanizomeon). This was tested in vitro using Human Bone Marrow CD34 and two progenitor cell lines KG1 and K562 Since (Aphanizomeon) can be derived from a simple aquatic algae, I feel it makes a strong argument, to use prior to Blood Derived Platelet Rich Plasma.

From our past studies, we regularly see Platelet levels around two million after centrifuge. We have not monitored these levels from fat extraction, But I feel confident there would be a similar increase in the circulating stem cells and growth factors deposited there.

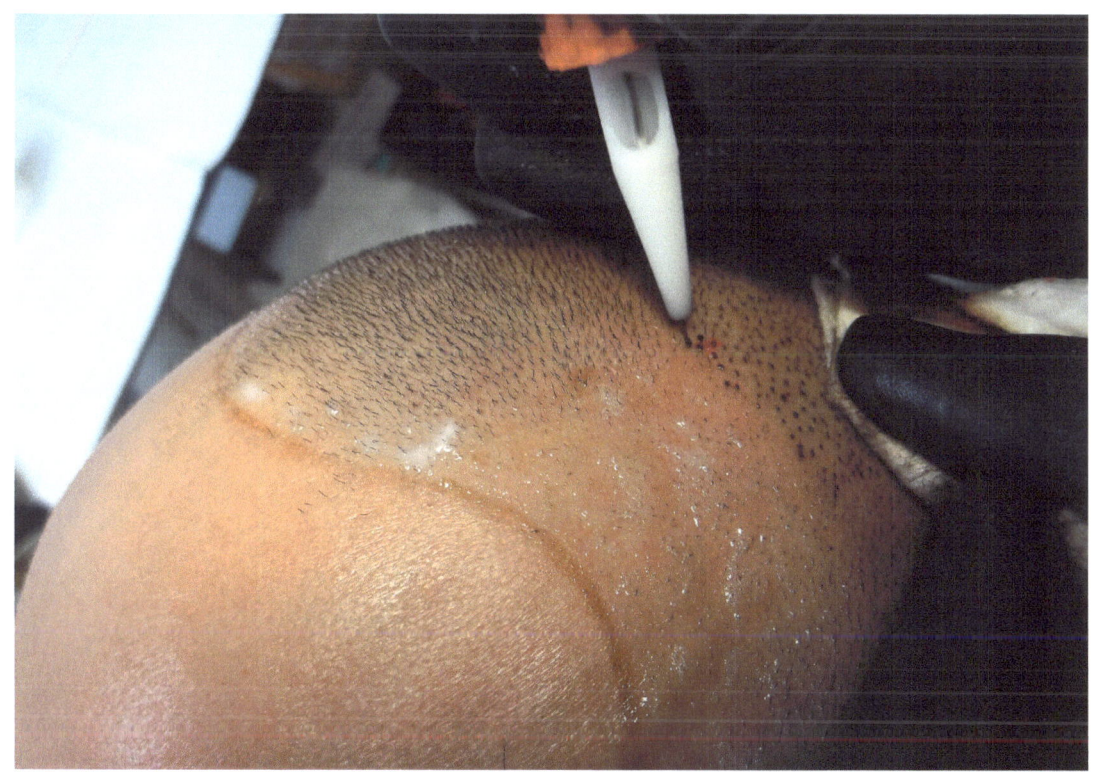

Injection process

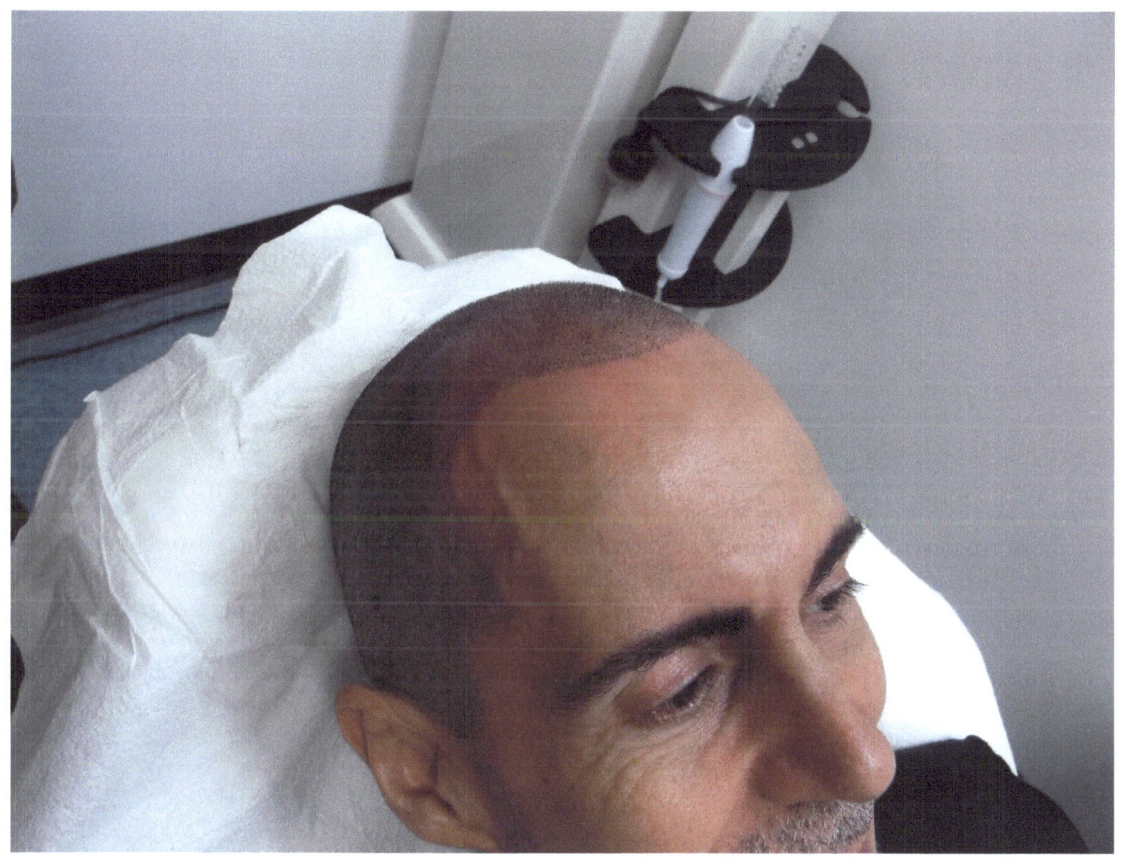

One week after

BEFORE

AFTER

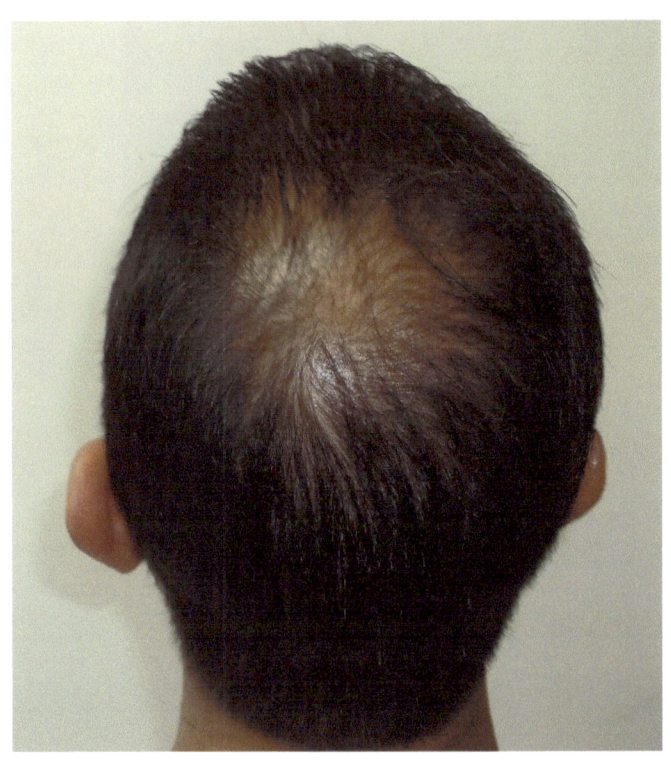
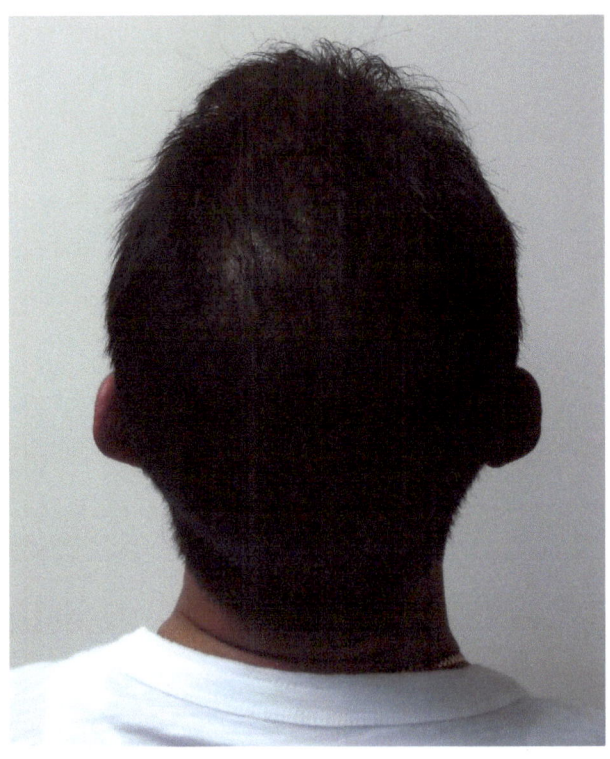
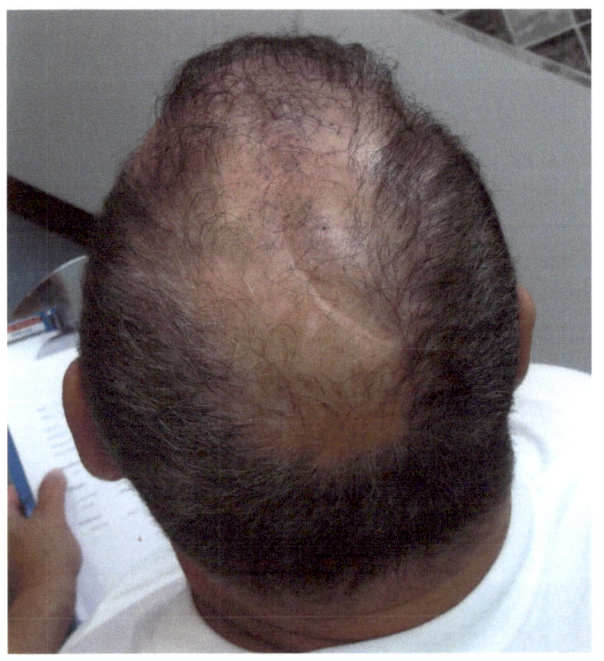
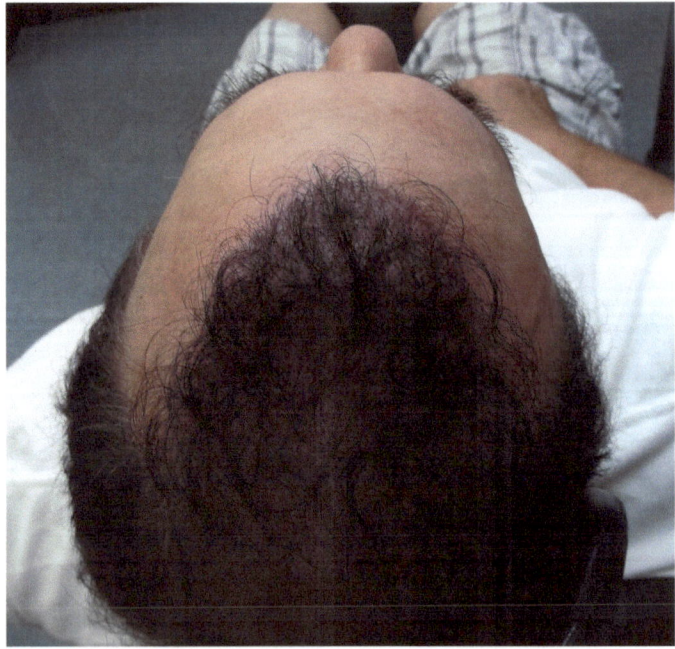

BEFORE

AFTER

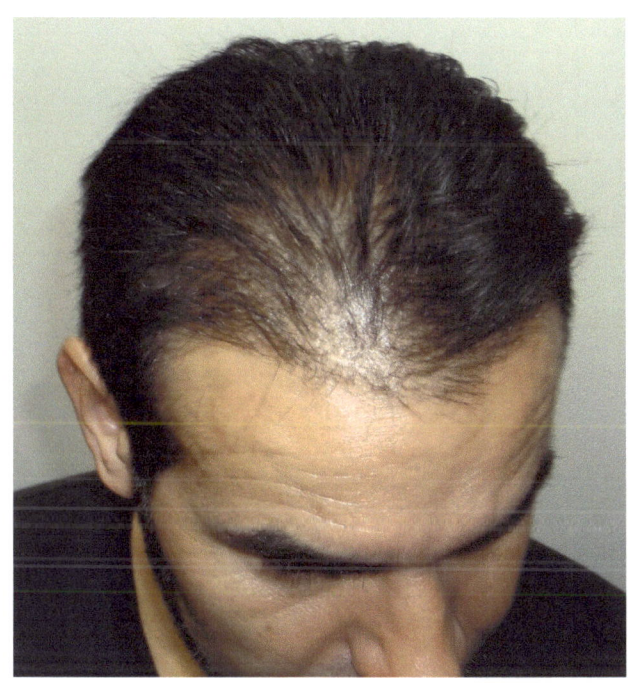
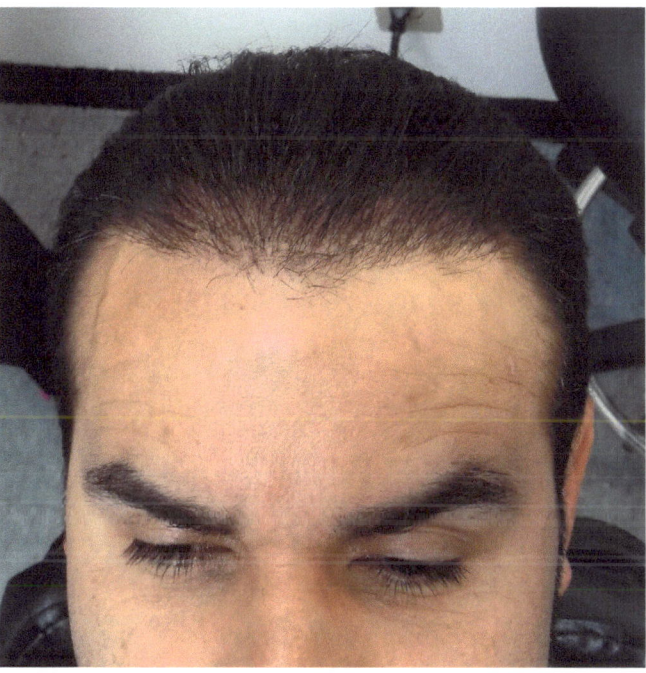

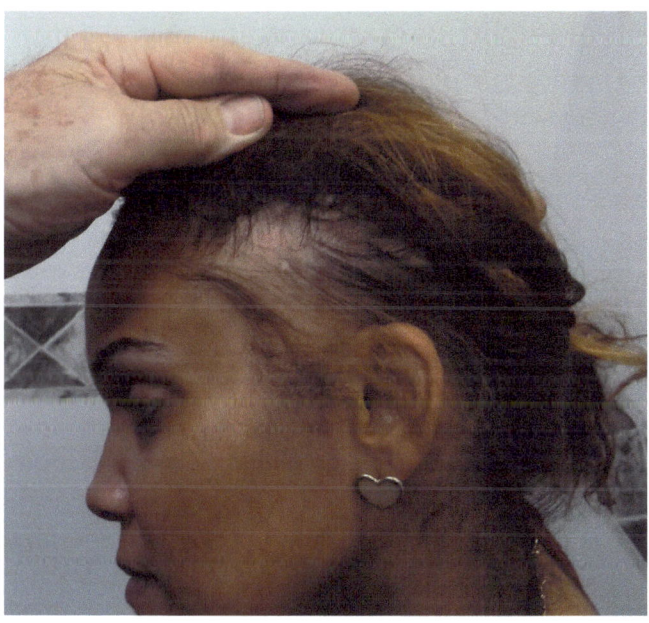
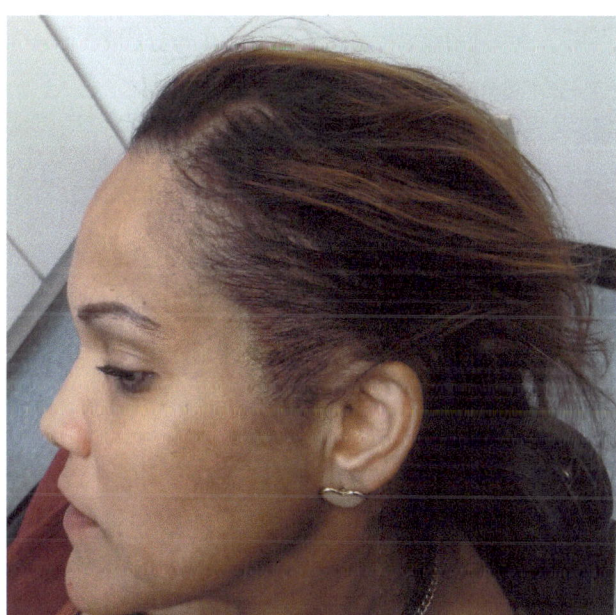

REFERENCES

2013 Cabot T. (March) what ever happened to stem cells Esquire – research –politics- stem cells

2011 Berkrot W. Krauskopf (stem cell in heart failure) Reuters. Article 2011

2008 Caplan A. treatment of human disease using Adult Meschenchymal stem cells NJCB files

2012 CBS News Could the next generation live to be 150? New York

2012 World Journal of Cardiology iv 312-326 Circulating endothelil & progenitor cells And long term exercise effects

2010 Drapeau MSc Cracking the Stem Cell Code Sutton Heart Press LLC

2012 The Holy Grail of Medicine on the Mystery and Power of Stem Cells, The Atlantic 1-19-2012

2007 Mobilization of CD34 +CD133 and CD34+CD133 stem cells in vitro related to Aphanizomenon modulation of CXCR4 exspression Cardiovascular Revascular Medcine 189 – 202 .

2011 Lin D. Parents Count on Extracting Stem Cells From Teeth NBC 4 Southern California

2011 Stem Cells Alternative to Knee Replacement ABC News Neporent L.

2011 Nocera R.MD. Cells that Heal Us From the Cradle to Grave. Scottsdale Multimedia inc.

2011 Paddock C. Reverse Diabetes with Stem Cells From Cord Blood. Medical News Today

2001 Orlic,Kajstura,Chimenti,Jakoniuk, Anderson Bone Marrow Cells regenerate Infarcted Myocardium Pub Med Gov.

2012 The Hindu Business line Stem Cell Market all Set To Grow.

2012 Young R. Stem Cell Use In the United States, New York Stem Cell Summit. P2

2007 J Nanosci Pub Med Human Plasma Protein Adsorption on Carbon-Based Materials